THE KOREA PATHFINDER LUNAR OBITER:

What you should know about Danuri (Enjoy the moon)

By Jason Howard

Table of contents

Chapter 1

THE LAUNCHING OF DANURI

South Korea successfully launched its first home-developed lunar orbiter becoming the seventh country to enter the highly fierce race to send spacecraft to the moon.

The Lunar Orbiter program was developed in response to the requirement to gather comprehensive images of potential Apollo landing sites.

South Korea's first trip to the moon took off at 7:08 p.m. EDT (2308 GMT) Thursday from Cape Canaveral Space Force Station in Florida. The launch was the second of the day from Florida's Space Coast, marking the fastest turnaround between launches at Cape Canaveral since 1967.

The Falcon 9 lifted out from Space Launch Complex 41 at Cape Canaveral carrying the Korea Pathfinder Lunar Orbiter, a 1,495-pound (678-kilogram) spacecraft the size of a large refrigerator that will gather data on lunar geology and look for signs of water ice buried in craters near the moon's poles.

The rocket's first stage landed on SpaceX's drone ship "Just Read the Instructions" parked downrange in the Atlantic Ocean east of Cape Canaveral.

The launch of the Falcon 9 rocket Thursday evening occurred 12 hours and 39 minutes after the liftoff of a United Launch Alliance Atlas 5 rocket with a U.S. military missile warning satellite at 6:29 a.m. EDT (1029 GMT) from pad 41, located about a mile-and-a-half (2.5 kilometers) north of SpaceX's launch complex.

"It is a very significant milestone in the history of Korean space exploration," Sang-Ryool Lee, president of KARI stated in a pre-recorded video. "If we are more determined and committed to technological development for space travel, we will be able to reach Mars, asteroids, and so on in the near future."

It will travel for four and a half months before entering lunar orbit in late December to begin its mission in January using six lunar payloads, vice science minister Oh Tae-Seog said in a briefing. The Korea Pathfinder Lunar Orbiter-KPLO is on a year-long mission, however, the period may be extended.

Danuri's missions include scouting for prospective landing sites as well as performing a wireless internet test by streaming a video of K-Pop group BTS's hit song **"Dynamite"**, according to South Korea's Electronics and Telecommunications Research Institute.
The launch comes after South Korea launched its own fully developed rocket **Nuri** in June that sent a test satellite into Earth's orbit. The country aims to devote finances in the following years to

launch an uncrewed spacecraft to the moon by 2031.

The mission represents a new chapter in the collaboration between the Korean and US space programs. The spacecraft is carrying NASA's ShadowCam, a camera co-developed by Arizona State University, and Malin Space Science Systems, an imaging firm located in San Diego. ShadowCam will take photographs of continuously shaded places, boosting the hunt for signs of ice deposits, according to a NASA statement.

This month is shaping up to be a busy one for NASA's lunar ambitions. NASA has scheduled an Aug. 29 launch date for Artemis I, an uncrewed spacecraft that will be the first in a series of missions planned to return astronauts to the moon for the first time since the Apollo program

The last time there was such a short period between two orbital-class rockets breaking off from Cape Canaveral was on Sept. 7 and 8, 1967, when a Thor Delta G rocket and an Atlas-Centaur rocket launched less than 10 hours apart. The Thor Delta G rocket launched a recoverable spacecraft named Biosatellite 2 with a series of biological research experiments, while the Atlas-Centaur transported NASA's Surveyor 5 lander to the moon.
A lunar mission made up the second half of Thursday's doubleheader, too.

The KPLO mission is a pathfinder, or predecessor, for South Korea's future objectives in space exploration, which include a robotic landing on the moon in the early 2030s. South Korea has also signed up to join the NASA-led Artemis Accords and might contribute to the U.S. space agency's human lunar exploration mission.

The KPLO mission is also called **Danuri**. The name Danuri is a combination of two Korean words "dal" which means moon and "nuri" which means enjoy, according to NASA Spaceflight, literally, Danuri means "enjoy the moon"

"The primary goal of this mission is technology development and demonstration," said Eunhyeuk Kim, the mission's project scientist from the Korea Aerospace Research Institute. "Also, utilizing the scientific equipment, we are expecting to collect some important data on the lunar surface."
The mission contains six scientific instruments and technology demonstration payloads.

KPLO will test a new South Korean spacecraft platform built for deep space missions, together with new communication, control, and navigation capabilities, including the certification of an "interplanetary internet" connection utilizing a disruption-tolerant network.

The Korea Lunar Pathfinder Orbiter is encapsulated within the Falcon 9 rocket's payload fairing.

The mission's scientific goals include mapping the lunar surface to aid identify future landing locations, assessing resources such as water ice on the moon, and studying the radiation environment near the moon.

The mission cost around $180 million to develop. The Falcon 9 rocket launched the KPLO spacecraft toward the moon on a low-energy, fuel-efficient ballistic lunar transfer trajectory, a path being pioneered by NASA's small CAPSTONE spacecraft, a tech demo mission that launched in June on a Rocket Lab mission and is scheduled to slip into orbit around the moon in November.

Instead of reaching the moon in a few days, like NASA's Apollo missions, KPLO will take roughly four months to complete the mission. KPLO's arrival date to the moon is determined on Dec. 16. The Falcon 9 put the spacecraft on a trajectory that would take it close to the L1 Lagrange point, a gravitationally-stable place roughly a million miles (1.5 million kilometers) from the sunlight side of the Earth, some four times further than the moon.

Gravitational forces will naturally drive the spacecraft back toward the Earth and the moon, where the Korean probe will be secured in orbit on Dec. 16. A sequence of propulsive movements with the spacecraft's engines will drive KPLO into a circular low-altitude orbit around 60 miles (100 kilometers) from the lunar surface by New Year's Eve.

After a month of commissioning and testing, the spacecraft's year-long main scientific mission should begin around Feb. 1. If the orbiter has adequate fuel, mission management might pursue an extended mission starting in 2024.
The launch Thursday marked SpaceX's 34th Falcon 9 mission of the year. The KPLO launch was also the 32nd space mission of the year from

Cape Canaveral to travel into orbit or toward more distant destinations.

The Korea Lunar Pathfinder Orbiter spacecraft during final testing in South Korea was stationed inside a firing room at a launch control center at Cape Canaveral, SpaceX's launch team began loading super-chilled, densified kerosene and liquid oxygen propellants into the 229-foot-tall (70-meter) Falcon 9 vehicle at T-minus 35 minutes.

Helium pressurant was also pumped into the rocket in the final half-hour of the countdown. In the last seven minutes before liftoff, the Falcon 9's Merlin main engines were thermally conditioned for flight in a technique known as "chilldown." The Falcon 9's guidance and range safety systems were also prepared for launch.

After liftoff, the Falcon 9 rocket vectored its 1.7 million pounds of torque — generated by nine

Merlin engines — to head east over the Atlantic Ocean.

The rocket surpassed the speed of sound in roughly one minute, then shut down its nine main engines two-and-a-half minutes after liftoff. The booster stage emerged from the Falcon 9's upper stage, then fired pulses from cold gas control thrusters and extended titanium grid fins to assist direct the spacecraft back into the atmosphere.

Two braking burns slowed the rocket from landing on the drone ship "Just Read the Instructions" roughly 400 miles (640 kilometers) downrange around nine minutes after liftoff.

The rocket flying on the KPLO mission, designated as B1052, launched on its sixth journey to space. It debuted as a side booster on two SpaceX Falcon Heavy missions in 2019, then SpaceX teams adapted it to fly as a Falcon 9 rocket.

While the first stage returned to Earth for landing, the Falcon 9's upper stage fired its single Merlin engine two times, first for six minutes to establish a parking orbit, then again at T+plus 34 minutes for a 60-second burn to launch the KPLO spacecraft on the way to the moon.

Separation of the KPLO spacecraft was verified at T+plus 40 minutes, 16 seconds. The probe was scheduled to extend its solar panels and begin transmitting to ground controllers after approximately 20 minutes of deployment from the Falcon 9, according to KARI.

THE MAIN GOALS OF DANURI

According to the NASA statement, the three main goals of the mission are:

- Realizing the first space exploration mission by the Republic of Korea (ROK).
- Developing and verifying space technologies suitable for deep-space exploration on future missions.
- Investigating the physical characteristics of the lunar surface to aid future robotic landing missions on the moon.

To achieve these goals, the spacecraft will carry a payload of five scientific instruments to include three cameras, a magnetometer, and a gamma-ray spectrometer. NASA is supplying one of the cameras, known as ShadowCam, which will be used to capture optical pictures at high-resolution of the permanently shadowed areas near the lunar poles of the Moon that are suspected to contain ice. ShadowCam's lead investigator is Mark Robinson, Professor of Geological Sciences at Arizona State

University's School of Earth and Space Exploration.

Danuri contains a magnetometer whose results may help scientists Better comprehend the residual magnetic field of the moon – In particular, hazy places where this field is exceptionally high.

Carey Institute officials said the photographs of Danuri would assist mission planners to locate excellent places for South Korea's future moon landing mission. And the ShadowCam – which is based on NASA's onboard LROC camera technology, but is more sensitive than its lunar survey spacecraft – Will look for ice in permanently darkened lunar craters.

It is thought that these holes hold a lot of ice.

Chapter 2

NASA's ROLE IN THE LAUNCHING OF DANURI

KARI and NASA initially discussed collaboration for the project in 2015 and signed a formal agreement to work together in December 2016.

Danuri is a joint mission between KARI and NASA with KARI managing the manufacturing and operation of the orbiter while NASA supports the mission with the development of one of the scientific payloads as well as aiding spacecraft communications and navigation, according to an agreement signed in 2016.

South Korea is also a party to the Artemis Agreements, a set of principles aimed to encourage ethical exploration of the Moon. South Korea Signed the accords in May 2021 becoming the ninth country to do so. Eleven more countries have followed suit since then.

NASA has shared experience in developing the missions and will offer access to the agency's Deep Space Network antennas across the globe to track the spacecraft.

"The KPLO Participating Scientist Program is an example of how international collaborations can leverage the talents of two space agencies, to achieve greater scientific and exploration success than individual missions," says Dr. Sang-Ryool Lee, the KPLO Project Manager, in a NASA statement.

"It's fantastic that the Korea Aerospace Research Institute (KARI) lunar mission has NASA as a partner in space exploration - we're excited to see the new knowledge and opportunities that will arise from the KPLO mission as well as from future joint KARI–NASA activities," Lee added.

NASA's role with KPLO goes beyond ShadowCam. The US space agency also picked nine researchers to participate in the mission

Chapter 3

FAST FACTS ABOUT DANURI

The following are fast facts about the Korea Pathfinder Lunar Orbiter -- known as Danuri in Korean -- which was launched aboard a SpaceX Falcon 9 rocket from Cape Canaveral Space Force Station in the U.S. state of Florida on Friday (Korean time) for South Korea's first lunar orbiter mimission.

- Danuri is a combination of the Korean words for moon and enjoy. It was proposed by a doctorate student in new materials engineering at the Korea Advanced Institute of Science and Technology and was chosen among 62,719 submissions in a national naming contest.

- Weighing 678 kilograms, Danuri is a cubic-shaped spacecraft with two solar arrays. For propulsion, it uses four high-powered thrusters and is also equipped with eight altitude control thrusters.

- Danuri marks South Korea's first space mission to travel beyond Earth's orbit.

- The Korea Aerospace Research Institute (KARI) project cost 236.7 billion won (US$180.5 million) and has had the participation of 40 private companies, 13 universities, and six government-funded research centers in South Korea.

- KARI first conceived the mission in 2013 as part of a long-term plan for space development. It is called a pathfinder to highlight South Korea's goal of advancing its space development, including a plan to send a lunar landing module to the moon in 2031.

- Danuri is South Korea's first lunar orbiter that was launched into space on Aug. 4, 2022, from Cape Canaveral Space Force

Station in the U.S. state of Florida.The spacecraft is scheduled to reach and start circling its designated orbit an altitude of 100 kilometers above the lunar surface -- in late December after traveling on a low-energy, fuel-efficient ballistic lunar transfer trajectory for 4 1/2 months.

- Danuri is carrying six scientific instruments, including an ultra-sensitive camera provided by the U.S. National Aeronautics and Space Administration (NASA), and will measure terrains, magnetic strengths, gamma rays, and other traits of the lunar surface during its yearlong mission. It will also identify potential landing sites for future lunar missions.

- A delay/disruption tolerant network instrument will be charged with testing whether a smooth internet connection between the Earth and the moon is possible. One of the pieces of content chosen to be digitally transferred is the BTS song "Dynamite," the first K-pop song to land at No. 1 on the U.S. Billboard Hot 100 chart.

THE KEY FACTS ABOUT DANURI

ROCKET: Falcon 9 (B1052.6)
PAYLOAD: Korea Pathfinder Lunar Orbiter
LAUNCH SITE: SLC-40, Cape Canaveral Space Force Station, Florida
LAUNCH DATE: Aug. 4, 2022
LAUNCH TIME: 7:08:48 p.m. EDT (2308:48 GMT)
Mass: 1,495 lbs (678 kilograms)
Target: Moon
Funding agency: Korea Aerospace Research Institute (KARI)
Notable firsts: South Korea's first lunar mission
WEATHER FORECAST: 80% chance of acceptable weather; Low risk of upper-level winds; Low risk of unfavorable conditions for booster recovery

BOOSTER RECOVERY: "Just Read the Instructions" drone ship east of Cape Canaveral

LAUNCH AZIMUTH: East

TARGET ORBIT: Ballistic lunar transfer trajectory

LAUNCH TIMELINE:

T+00:00: Liftoff

T+01:12: Maximum aerodynamic pressure (Max-Q)

T+02:31: First stage main engine cutoff (MECO)

T+02:34: Stage Separation

T+02:42: Second stage engine start (SES 1)

T+03:15: Fairing jettison

T+06:49: First stage entry burn ignition (three engines)

T+07:19: First stage entry burn cutoff

T+08:33: Second stage engine cutoff (SECO 1)

T+09:01: First stage landing

T+34:15: Second stage engine start (SES 2)

T+35:15: Second stage engine cutoff (SECO 2)

T+40:16: KPLO spacecraft separation

MISSION STATS:

168th launch of a Falcon 9 rocket since 2010

176th launch of Falcon rocket family since 2006

6th launch of Falcon 9 booster B1052

145th Falcon 9 launch from Florida's Space Coast

93rd Falcon 9 launch from pad 40

148th launch overall from pad 40

110th flight of a reused Falcon 9 booster

3rd dedicated SpaceX launch for South Korean customer

2nd SpaceX launch with a lunar payload

34th Falcon 9 launch of 2022

34th launch by SpaceX in 2022

34th orbital launch attempt based out of Cape Canaveral in 2022.